Genius Book

100+ smart facts that will make you say “I didn’t know that”

Top Tier Education Program

ISBN 9791224314325

DEDICATION

To those who enjoy asking why things work the way they do, even when the answer has no immediate use. This book is for curious minds who believe that understanding small details about the world can quietly change how we see it every day.

CONTENTS

ACKNOWLEDGMENTS i

INTRODUCTION 1
A SHORT TALK WITH THE READER Pg 5
100+ SMART FACTS Pg 16
CONCLUSIONS Pg 129
ABOUT THE AUTHOR Pg 131

ACKNOWLEDGMENTS

This book draws on the work of scientists, researchers and thinkers whose curiosity helped uncover how the world actually works. Their insights made these pages possible. Thanks also to readers who remain open to learning unexpected things, even when they seem trivial at first glance.

INTRODUCTION

There are books you read from beginning to end, carefully turning each page as if you were following a path someone else has already mapped out for you. This is not one of those books. This is the kind of book you can open at random while waiting for your coffee, brushing your teeth, sitting on the couch, or simply trying to win an argument with a friend using a very specific and completely unexpected fact. Inside these pages you will find ideas, discoveries, strange truths and small scientific wonders that can be read in any order, in any mood, and at any time. You don't need a plan. You don't need a schedule. Just curiosity. And maybe two minutes.

Why Your Brain Loves Random Facts More Than You Think

Have you ever noticed how easy it is to remember something completely useless, and how strangely difficult it is to recall something important, like where you left your keys five minutes ago? You might forget a password you use every day, but you can remember that octopuses have three hearts or that honey never spoils. It feels random. It feels irrational. But it's actually not.

Your brain is not designed to remember everything. If it were, you would be overwhelmed within minutes. Instead, it's designed to remember what is interesting, surprising, emotionally charged or slightly unusual. In other words, your brain is built to notice things that break patterns.

Most of what you experience every day is predictable. The same streets, the same sounds, the same apps, the same routines. Your brain learns to ignore a lot of this because it doesn't need to waste energy processing things that are already familiar. This is called habituation, and it's the reason why you stop noticing the sound of a fan after a few minutes or the feeling of your clothes on your skin.

But the moment something strange appears, your brain wakes up.

A surprising fact is a small violation of expectation. It's a tiny glitch in your mental model of how the world works. And your brain loves glitches, because glitches might mean danger, opportunity or change. Thousands of years ago, noticing something unusual could mean the difference between eating and being eaten. A movement in the grass that didn't fit the pattern might be the wind. Or it might be a predator.

So your brain evolved to pay special attention to the unexpected.

Today, instead of predators, we notice weird facts. The mechanism is the same. When you learn that bananas are technically radioactive or that sharks existed before trees, your brain experiences a small cognitive surprise. It goes: wait, that doesn't sound right. And in that moment, your attention spikes.

Attention is the gateway to memory.

The more surprised you are, the more likely your brain is to tag that information as worth storing. It releases a small dose of dopamine, the same neurotransmitter associated with reward and motivation. That chemical signal tells your brain: this is interesting, keep it.

This is why random facts stick.

They don't feel like work. They feel like discovery. They create a tiny moment of wonder, and wonder is incredibly efficient at opening the door to learning. You may not remember a list of dates from a history class, but you might remember that Cleopatra lived closer in time to the invention of the iPhone than to the building of the pyramids.

Not because it's useful.

But because it's unexpected.

And every time you encounter something unexpected, your brain quietly thinks: this might matter someday.

Why Knowing "Useless" Things Is Actually Useful

We tend to divide knowledge into two categories: useful and useless. Useful knowledge helps you pass exams, fix problems, get jobs, make decisions and survive everyday life. Useless knowledge is everything else. The weird stuff. The fun facts. The trivia. The kind of information that seems to have no clear application beyond winning a quiz night or filling an awkward silence during dinner.

But this division is not as accurate as it feels.

Your brain doesn't really store information in tidy little folders labeled "important" and "unimportant." Instead, it builds networks. Everything you learn becomes part of a larger web of associations, where ideas connect in ways that are often invisible until the moment you need them. A random fact about how Velcro was inspired by burrs sticking to dog fur might not seem practical today, but it reinforces a deeper pattern: innovation often comes from observing nature.

That pattern might show up later when you're solving a problem, designing something, or trying to understand how a system works.

Many breakthroughs don't come from deep specialization alone. They come from connecting two ideas that were never meant to meet. Something from biology meets something from engineering. Something from physics meets something from art. Something from psychology meets something from marketing. And suddenly, there's a new solution.

This kind of thinking is called associative thinking.

And associative thinking feeds on variety.

The more different pieces of information you collect, even if they seem trivial on their own, the more raw material your brain has to work with when it needs to generate new ideas. Think of your mind as a toolbox. If you only ever collect one type of tool, you'll solve problems in the same way over and over again. But if your toolbox contains strange, unexpected instruments, you might find creative shortcuts that others miss.

Random facts expand your mental toolkit.

They also improve your ability to understand conversations, articles and real-world situations. Many things make more sense when you have a wide background of general knowledge. Technologies, trends, scientific discoveries and cultural changes often rely on ideas that cross disciplines. Knowing a little about many things can help you follow connections that are otherwise invisible.

And there's another benefit.

Sharing small pieces of surprising information is one of the simplest ways humans connect with each other. Conversations often begin with curiosity. "Did you know that...?" is a social bridge. It signals interest, playfulness and openness. It invites discussion. It makes interactions more engaging.

In that sense, knowing so-called useless facts is not useless at all.

It helps you think differently.

It helps you understand more.

And sometimes, it simply helps you start a conversation that wouldn't have happened otherwise.

The Power of Reading One Page at a Time

There is a quiet pressure that comes with most books. You are supposed to read them in order. You are supposed to follow the argument from beginning to end. You are supposed to remember what came before in order to understand what comes next. Even when nobody says it out loud, it feels like there is a correct way to read.

This book does not require that kind of commitment.

In fact, it works best when you don't treat it as a continuous narrative. You can open it anywhere. You can read a single page and close it again. You can jump from one section to another without losing anything along the way. Each fact is designed to stand on its own, like a small window into a bigger world.

This matters more than it seems.

Modern life is full of interruptions. Notifications, messages, calls, deadlines, conversations, updates and endless streams of information compete for your attention. Long, uninterrupted stretches of reading time are rare. And when they do appear, they are often already claimed by other priorities.

Short, self-contained reading experiences fit into the spaces that longer books cannot reach.

Two minutes while waiting in line. Three minutes before a meeting starts. A few moments before going to sleep. These small gaps might not seem important, but they add up. Over time, reading one page here and one page there can expose you to dozens of new ideas without requiring any special effort or planning.

It also changes how you approach learning.

When there is no pressure to finish a chapter or remember a complex argument, curiosity becomes the main driver. You read because you want to know what happens next, not because you feel obligated to keep going. And curiosity is one of the strongest motivations for learning that exists.

There is also a psychological effect at play.

Completing a small unit of information gives your brain a sense of closure. You start something and you finish it within minutes. That small completion releases a feeling of progress, even if the task itself was simple. Over time, these tiny moments of completion reinforce the habit of coming back for more.

One page becomes two.

Two pages become ten.

Ten pages become a collection of ideas that slowly reshape how you see the world.

You don't need long sessions.

You don't need silence or perfect concentration.

Sometimes, all it takes is a single page.

A SHORT TALK WITH THE READER

Before you go any further, let's agree on something important: you don't owe this book anything. You don't have to finish it. You don't have to read it in order. You don't even have to remember most of what's inside. This is not a test, and there are no bonus points for reaching the last page. Think of this as a collection of small, portable ideas you can carry with you whenever you feel like it. Some pages might make you pause, some might make you smile, and some might make you look at everyday things in a slightly different way. Read one fact. Read five. Skip ten. Come back later. This book will still be here, exactly where you left it.

How to Read This Book the "Wrong" Way

Most books come with an invisible set of instructions.

Start at page one. Continue in order. Pay attention. Try to remember what you read because it might be important later. Don't skip ahead. Don't jump around. Don't miss anything.

Even when nobody says this out loud, it's the rule we've learned since childhood. Books are supposed to be read from beginning to end. That's how stories work. That's how arguments work. That's how textbooks work. There is a path, and you are expected to follow it.

But this book was not designed that way.

In fact, the "right" way to read it is very likely the wrong one.

If you sit down and try to read it from the first fact to the last in a single, determined sequence, you might accidentally turn it into something it was never meant to be: a task. A project. Something to complete. And the moment reading becomes something to complete, your brain quietly moves it into the same mental category as laundry, emails or assembling furniture.

Your attention changes.

Instead of curiosity, you feel obligation. Instead of interest, you feel effort. Instead of discovery, you feel progress tracking. You might find yourself checking how many pages are left, wondering how long it will take to finish, and asking yourself whether you should continue now or later.

That's not what this is for.

This book is closer to a conversation than to a lecture. It doesn't need continuity. It doesn't need a narrative arc. There is no build-up you have to follow and no grand conclusion waiting at the end that will suddenly make everything else make sense. Each page is a self-contained moment. A small mental detour. A brief interruption in whatever you were doing before you opened it.

And interruptions can be useful.

Imagine opening the book in the middle and landing on a fact you weren't expecting. Something about how the GPS system needs to account for relativity in order to work correctly, or how the smell of rain has a specific scientific name, or how some animals can see parts of the light spectrum that are completely invisible to you.

You didn't plan to learn that.

You weren't studying for anything.

But now you know it.

And more importantly, now you see something familiar in a slightly different way. The next time it rains, you might think about the chemistry in the air. The next time you use your phone for directions, you might briefly remember that satellites have to adjust for time dilation. The world becomes just a little more detailed, even if nothing around you has changed.

This is the effect of reading in fragments.

When you read a traditional book in order, you are following someone else's structure. You are moving through a path that was carefully arranged for coherence and flow. That kind of reading is valuable, but it requires time, attention and a willingness to stay inside the same mental context for a while.

Fragmented reading is different.

It allows you to step in and out of learning without needing to rebuild the entire context every time. You don't have to remember what came before, because nothing depends on it. You don't have to prepare, because each page brings its own introduction. You don't even have to be in the right mood.

You can read when you are tired.

You can read when you are distracted.

You can read when you have exactly ninety seconds to spare.

And when you close the book, you are not in the middle of something unfinished. You are done. That page is complete. The experience has a beginning and an end, even if it lasted less than two minutes.

There is also a small psychological advantage in doing things the "wrong" way.

When you give yourself permission to skip, jump, browse and explore, you remove the pressure that often comes with doing things properly. You are no longer trying to be a good reader. You are simply being a curious one. And curiosity tends to show up more easily when there are no rules to follow.

Think about how you use the internet.

You rarely start at the first page of a website and move systematically through every link in order. You click on something that catches your attention. Then you follow another link. Then another. Sometimes you end up somewhere completely unrelated to where you started, but along the way you pick up ideas, insights and small pieces of information that might stay with you for years.

This book works best when you treat it the same way.

Open it at random.

Let the page choose you.

Follow whatever seems interesting in the moment.

If you read the book from start to finish, you will still find interesting things inside. But if you read it out of order, dipping in and out whenever you feel like it, you may find that it fits more naturally into your day. It becomes less like an activity you schedule and more like a resource you reach for.

A few minutes in the morning.

A single page before bed.

One fact while waiting for someone to arrive.

Over time, these small encounters accumulate.

You might not notice it immediately, but your mental model of the world starts to expand. Everyday objects become slightly more interesting because you know something about how they came to be. Ordinary phenomena become less ordinary because you have seen the mechanism behind them. Conversations become richer because you have new ideas to share.

All without ever sitting down for a long study session.

All without trying to read anything the "right" way.

So feel free to ignore the usual rules.

Skip pages.

Jump ahead.

Close the book halfway through and come back days later.

If anything, that's exactly what you're supposed to do.

Why You Only Need Two Minutes

There is a very common reason why people don't read as much as they would like to, and it has nothing to do with intelligence, motivation or interest.

It's time.

Or more precisely, the feeling that reading requires a certain kind of time. Quiet time. Uninterrupted time. The kind of time you imagine having on a long train journey or on a slow Sunday afternoon when there's nothing else to do. Reading is often associated with comfort, stillness and focus. A cup of tea. A couch. Silence.

And while that image is pleasant, it's also slightly misleading.

Because if reading always required that kind of setting, most people would rarely do it at all. Modern life is not built around long stretches of calm attention. It's built around interruptions. Messages arrive. Notifications appear. Meetings start earlier than expected. Something needs your response right now. Even when you do find a moment to sit down, it often feels temporary, as if something else might demand your attention at any second.

So reading becomes something you postpone.

You tell yourself that you will do it later, when you have more time. Later tonight. This weekend. On vacation. But when that future moment arrives, it is usually filled with something else. Rest. Work. Social plans. Another book you feel you should be reading instead. And so the intention to read quietly moves further into the background.

Not because you don't want to read.

But because you think you need more time than you actually do.

This is where short, self-contained reading changes the equation.

A single page can be read in about two minutes. Sometimes less. That's roughly the amount of time it takes to wait for an elevator, stand in line at a coffee shop, sit through a traffic light cycle, or wait for a video to finish buffering. These are small, almost invisible moments that occur throughout the day. They don't feel like opportunities for learning because they are not traditionally framed that way.

But they are still time.

And more importantly, they are time that usually goes unused.

You check your phone.

You scroll through something you won't remember later.

You reread the same message thread out of habit.

None of this is wrong, of course. But it does mean that your brain is already accustomed to filling these small gaps with information. The difference is that most of the information you encounter in these moments disappears quickly. It is designed to capture your attention, not to stay with you.

Reading a single fact works differently.

It offers closure. You begin, you read, and you finish within a short, defined period. There is no cliffhanger. No need to keep going. No sense that you are in the middle of something unfinished. This matters, because unfinished tasks tend to create a mental loop. Your brain keeps them active in the background, which can make it harder to focus on other things.

A completed unit of reading does the opposite.

It provides a clear start and a clear end. That small sense of completion can feel surprisingly satisfying, even if the task itself was simple. Over time, this creates a subtle shift in how you perceive reading. Instead of being a large, demanding activity that competes with everything else, it becomes something that can fit into the natural pauses of your day.

And once something fits easily into your routine, it is more likely to become a habit.

Habits do not form because something is important. They form because something is easy to repeat. The less effort an activity requires, the more often you are willing to do it again. If reading requires a quiet room and an hour of uninterrupted time, you may only do it occasionally. But if reading requires two minutes and no preparation, you might do it without even thinking about it.

Two minutes in the morning while you wait for your coffee to cool down.

Two minutes in the afternoon between tasks.

Two minutes in the evening before turning off the light.

None of these moments feel significant on their own. But together, they create a steady rhythm of exposure to new ideas.

Over the course of a week, those two-minute sessions might add up to twenty or thirty pages. Over a month, they might introduce you to dozens of facts you didn't know before. And because each one is surprising in some small way, they are more likely to stay with you than a longer, more abstract explanation would.

There is also an important emotional component.

When an activity feels manageable, you are more willing to start it.

When you know you only need two minutes, there is very little resistance. You are not committing to a long session. You are not promising yourself that you will finish a chapter. You are simply opening the book, reading a page, and closing it again.

That small action lowers the barrier to entry.

And once the barrier is low enough, curiosity can do the rest.

Sometimes you will stop after one page.

Sometimes you will read three or four.

Sometimes you will forget about the time entirely and keep going.

But you never had to plan for it.

You never needed the perfect moment.

All you needed was two minutes.

The Power of Saying "Did You Know That...?"

There is a very simple sentence that has started more conversations than almost any other in human history.

"Did you know that...?"

It doesn't sound impressive. It's not poetic. It's not clever. But it has a very specific effect. The moment someone hears it, their attention shifts. Even if they weren't interested a second ago, they are now. Even if they were about to check their phone or think about something else, they pause, just for a moment, to see what comes next.

Because that sentence promises a surprise.

It suggests that there is a small gap between what you know and what you are about to know. And humans are naturally drawn to closing gaps like that. Psychologists sometimes refer to this as an information gap. When you become aware that there is something you don't know, especially if it seems interesting or unexpected, your brain becomes motivated to fill in the missing piece.

"Did you know that sharks existed before trees?"

"Did you know that octopuses have three hearts?"

"Did you know that there is a cloud that smells like vanilla in space?"

Even if you had no prior interest in marine biology or astronomy, you are now curious. And curiosity has a way of pulling you forward.

But what makes this sentence especially powerful is not just its ability to capture attention. It's what happens after the fact is shared.

Something changes in the social dynamic.

You are no longer just exchanging opinions or reacting to events. You are sharing a small piece of the world. A detail. A mechanism. A strange truth that makes reality slightly more interesting than it appeared a moment ago. And that shared moment of surprise creates a connection.

Think about how often conversations become repetitive.

You talk about the weather.

You talk about work.

You talk about plans for the weekend.

None of this is bad, but it is predictable. It follows familiar patterns. A surprising fact, on the other hand, interrupts that pattern in a gentle way. It introduces novelty. It invites a reaction that is not scripted.

"What? That can't be true."

"Really?"

"I had no idea."

Suddenly, the conversation has energy again.

And from that single fact, other ideas can emerge. Someone might ask a follow-up question. Someone else might share something related. You might find yourselves discussing how GPS works, why certain animals behave the way they do, or how a particular invention came to exist.

All because of one unexpected detail.

Sharing knowledge in this way also signals something subtle about the person who shares it. It suggests curiosity. Engagement. A willingness to notice things that others might overlook. It's not about showing off or proving expertise. In fact, the most enjoyable facts are often the ones that seem too strange to be useful.

That's what makes them fun.

And fun is contagious.

When someone shares something surprising, it often encourages others to do the same. The conversation becomes a chain of discoveries. One fact leads to another, which leads to a story, which leads to a question. Before you know it, you have moved far beyond the original topic.

This kind of exchange is different from debating or persuading.

It's collaborative.

You are not trying to win an argument. You are exploring together. You are adding small pieces to a shared picture of how the world works. And because the stakes are low, there is no pressure to be right or to reach a conclusion. You are simply curious.

There is also a memory benefit.

When you hear a surprising fact in conversation, especially one that made you laugh or pause, you are more likely to remember it later. Social context strengthens memory. The fact becomes associated with a moment, a place, a person. The next time you think about it, you might also recall where you were when you first heard it.

And then you might share it again.

This is how ideas spread.

Not through lectures or formal presentations, but through small, informal exchanges. A sentence here. A detail there. A brief moment of surprise that travels from one person to another, reshaping their understanding just a little.

So the next time you open this book and find something that makes you pause, consider sharing it.

Start with four simple words.

"Did you know that...?"

What Happens When You Open a Page at Random

There is something strangely satisfying about randomness.

We experience it all the time, often without noticing. A shuffled playlist that plays a song you haven't heard in years. A recommendation that appears out of nowhere and turns out to be exactly what you were in the mood for. A conversation that unexpectedly drifts into a topic you didn't know you cared about.

Randomness interrupts routine.

And routine, while useful, has a way of making everything feel predictable. The same information flows past you every day. The same types of headlines. The same updates. The same conversations about the same topics. Your brain learns what to expect, and once it knows what to expect, it begins to filter aggressively. Anything that doesn't immediately seem relevant gets ignored.

This is efficient.

But it also means that many interesting ideas never get through.

When you choose what to read next based only on what you already know you like, you tend to reinforce the same mental pathways. You read more of the same topics, in the same style, from the same perspectives. Over time, your knowledge becomes deeper in certain areas, but also narrower.

Opening a page at random works differently.

It bypasses your usual preferences.

You don't decide what you want to learn. You allow chance to decide for you. And because of that, you might encounter something you would never have searched for intentionally. A fact about how certain metals remember their shape. A detail about why time moves slightly faster at higher altitudes. A strange story about an invention that happened by accident.

None of these were on your list.

But now they are in your mind.

Random exposure creates unexpected connections. Something you read today might not seem relevant to anything you are doing right now. But weeks later, in a completely different context, that piece of information might suddenly become useful. You might recall it during a conversation, while solving a problem, or when trying to understand how a new technology works.

The connection often happens silently.

You don't always notice when your brain links two ideas together. But when it does, the result can feel like intuition. You have a sense that something is related, even if you can't immediately explain why. That sense often comes from prior exposure to small pieces of information that were stored away without any obvious purpose.

Random reading increases the chances of these moments.

It introduces variety into your mental environment. It disrupts the pattern of consuming only what you already expect to consume. And because each page in this book stands alone, there is no penalty for skipping around. You are not missing a chapter that will be needed later. You are not losing the thread of an argument.

You are simply encountering a new idea.

There is also an element of play.

When you open the book at random, you are turning reading into a small game. What will you find this time? Something about space? Something about the human body? Something about a material you use every day without thinking about it? The uncertainty adds a layer of interest that a fixed sequence cannot provide.

Playfulness makes learning feel lighter.

And when learning feels light, you are more likely to return to it. Not because you have to, but because you are curious about what might come next. Even if "next" is determined by chance.

Over time, this habit of exploring without a plan can influence how you approach information outside of the book as well. You might become more open to reading articles that fall outside your usual interests. You might follow links that lead to unfamiliar topics. You might notice patterns in areas you previously ignored.

All because you got used to opening a page without knowing what you would find.

Randomness does not guarantee relevance.

But it greatly increases the range of possibilities.

And sometimes, that is exactly what your brain needs.

Why Curiosity Is More Contagious Than You Think

Curiosity is often treated as a personal trait.

Some people are curious. Others are not. Some people ask questions all the time. Others prefer to stick with what they already know. It feels like something internal, almost fixed, as if you either have it or you don't.

But curiosity doesn't always work that way.

In many cases, curiosity is triggered by the people and the environment around you. It spreads quietly through conversations, shared interests and small moments of surprise. When someone near you becomes interested in something, it subtly changes your own level of attention. You start to notice what they notice. You begin to wonder what they wonder about.

This is why enthusiasm can be infectious.

Think about the last time someone explained something to you with genuine excitement. It might have been about a new technology, a scientific discovery, a strange historical event or even a hobby you had never considered before. At first, you might not have cared much about the topic itself. But their interest made you listen.

And as you listened, your own curiosity grew.

Not because the subject suddenly became important in an objective sense, but because it was presented in a way that suggested there was something worth noticing. Curiosity often begins with a simple signal: this is interesting. That signal doesn't have to come from within. It can come from someone else.

Small facts are particularly good at sending that signal.

They are easy to share. They don't require long explanations or deep background knowledge. They can be delivered in a sentence or two, and yet they carry enough surprise to spark a reaction. When someone hears that there are more possible games of chess than atoms in the observable universe, or that certain materials can heal themselves after being damaged, it creates a brief pause.

That pause is curiosity in action.

During that moment, the listener's mental model of the world is slightly disrupted. Something they assumed was simple becomes more complex. Something they thought they understood now seems incomplete. And that gap creates an opening for further interest.

Curiosity moves through social networks in subtle ways.

One person learns something new and shares it with two others. Those two might share it with someone else, adding their own interpretation or an additional detail. Over time, a single piece of information can travel far beyond its original source, not because it was part of a formal lesson, but because it was interesting enough to mention.

This is how many ideas spread in everyday life.

Not through structured presentations, but through informal exchanges. A surprising statistic mentioned in passing. A story about an invention that happened by accident. A fact about how a common object works that changes the way someone sees it.

These moments accumulate.

If you are regularly exposed to curious people, you are more likely to become curious yourself. You begin to expect that there is always something more to learn. That everyday objects have hidden mechanisms. That familiar processes have deeper explanations. The world starts to feel less static and more dynamic.

And once that shift happens, it becomes self-reinforcing.

You start to look for interesting details on your own. You pay attention to things you might have ignored before. You ask more questions. You notice patterns. Curiosity becomes less of a reaction and more of a habit.

Sharing what you learn accelerates this process.

When you tell someone about a surprising fact, you are not only transferring information. You are also modeling curiosity. You are showing that it is normal to be interested in how things work, why they exist, or what makes them unusual. That attitude can encourage others to adopt the same mindset.

And when they do, they might share something back.

A new detail.

A different perspective.

Another small mystery.

Bit by bit, curiosity moves through groups, conversations and communities. It doesn't need formal structures or scheduled lessons. It only needs moments of attention and a willingness to be surprised.

So the next time you find yourself wondering about something unexpected, remember that curiosity rarely stays in one place for long.

It travels.

The Secret Pleasure of Learning Without Trying

There is a moment, sometimes very brief, when you realize that you have learned something without intending to.

You didn't sit down to study.

You didn't take notes.

You didn't prepare yourself to remember anything.

And yet, later in the day, you catch yourself thinking about a fact you came across almost by accident. Maybe it was something about how your brain processes time, or why certain materials feel colder than others even when they are at the same temperature. Maybe it was a detail about space, or evolution, or how a familiar technology actually works.

You weren't trying to learn it.

But you did.

This kind of learning often feels different from the kind you associate with effort. It arrives quietly, without the tension that usually comes with concentrating on something important. It doesn't demand your full attention. It doesn't require you to memorize anything or prepare for a test. It simply happens as a byproduct of curiosity.

And that can make it more enjoyable.

When learning is framed as an obligation, it often comes with resistance. You know you should pay attention. You know you should understand the material. You might even feel pressure to remember it for later. All of this can make the process feel heavier than it needs to be.

But when learning happens incidentally, the experience changes.

You encounter an idea because it is interesting, not because it is required. You spend a few seconds thinking about it, not because you have to, but because you want to. And in that moment, your brain is more relaxed, more open to forming connections.

Relaxed attention can be surprisingly effective.

Research in psychology suggests that information encountered in low-pressure environments can still be retained, especially when it is distinctive or emotionally engaging. A surprising fact, a clever observation or an unexpected comparison can create a small mental imprint, even if you only read it once.

Over time, these small imprints accumulate.

You might not notice the process while it's happening. But gradually, your general understanding of how things work begins to expand. You have more reference points. More analogies. More examples to draw from when you are trying to make sense of something new.

This is sometimes described as building mental models.

Mental models are simplified representations of how the world operates. They help you predict outcomes, understand systems and interpret events. The more varied your mental models are, the more flexible your thinking becomes. You can approach problems from different angles. You can recognize patterns that others might miss.

And many of these models begin with simple observations.

A fact about how ants organize themselves might later influence how you think about teamwork. A detail about how light travels might change the way you understand communication technologies. A story about an accidental invention might encourage you to experiment more freely when you are working on something new.

None of these connections are planned.

They emerge over time, as your brain links ideas together in the background.

Learning without trying also removes the fear of getting things wrong.

When there is no test at the end, there is no penalty for misunderstanding something at first. You can revisit the idea later, perhaps from a different perspective. You can refine your understanding gradually, without pressure. This makes the process feel less like a performance and more like exploration.

Exploration invites experimentation.

You become more willing to follow a question, to look up an unfamiliar term, or to ask why something works the way it does. And each small step adds another layer to your understanding.

There is also a sense of quiet satisfaction that comes from realizing you know something new.

You might notice it during a conversation, when you recognize a reference that would have passed you by before. You might see it when you encounter a news article that mentions a concept you have already come across in a simpler form. Suddenly, the information feels more accessible.

You are not starting from zero.

And that can make complex topics seem less intimidating.

Over time, learning without trying can change your relationship with knowledge itself. It becomes less about effort and more about discovery. Less about obligation and more about interest. You begin to see learning not as a separate activity, but as something that happens naturally when you engage with the world.

Sometimes all it takes is a single page.

A brief pause.

A moment of curiosity.

And before you realize it, you have learned something new.

100+ SMART FACTS

PART I
PHYSICS OF EVERYDAY REALITY

FACT 1

COFFEE STAINS DRAW THEIR OWN RINGS

When a droplet's edge "pins," evaporation is strongest at the rim. To replace lost liquid, capillary flow carries suspended particles outward, leaving a dense perimeter ring as it dries.

FACT 2

TWISTING SPAGHETTI TAMES CRACK CASCADES

Dry spaghetti usually fractures into many pieces because the first break launches bending waves that trigger secondary breaks. Twisting stores torsion that is rapidly released, reducing the cascade and enabling clean two-piece breaks.

FACT 3

STRETCHING RUBBER TURNS ENTROPY INTO HEAT

Rubber elasticity is largely entropic: stretching aligns polymer chains, reducing their entropy. In fast stretches this entropy change appears as a measurable temperature rise; releasing tension can cool the rubber momentarily below ambient.

FACT 4

HOT WATER CAN FREEZE FIRST SOMETIMES

Under specific conditions, initially hotter water can reach freezing sooner because evaporation, convection, dissolved gas content, and supercooling can differ between samples. The effect is real in some setups but not universal.

FACT 5

BANANA PEELS CREATE A LUBRICATING GEL

A banana peel isn't "mysteriously slippery." Under pressure, its surface can release a polysaccharide-rich gel that dramatically lowers friction against surfaces like flooring, acting like a thin natural lubricant layer.

FACT 6

SOME BEER BUBBLES SINK BY CIRCULATION

In stout, small bubbles plus the glass shape can create a circulation: fluid moves downward near the wall and upward in the center. Bubbles get swept down near the edges, appearing to “sink” despite buoyancy.

FACT 7

FALLING MAGNETS BRAKE THEMSELVES IN COPPER

A moving magnet induces eddy currents in a conducting tube. Those currents create magnetic fields that oppose the motion (Lenz's law), converting gravitational potential energy into heat and producing a slow, steady terminal fall.

FACT 8

CORNSTARCH TURNS SOLID WHEN SHEARED HARD

Dense cornstarch–water suspensions can “jam” under sudden stress: particles form force-bearing networks, sharply increasing viscosity (shear thickening). That’s why you can run on it briefly but slowly sink at rest.

FACT 9

SOAP PROPULSION COMES FROM SURFACE TENSION

Soap lowers surface tension where it spreads. That creates a surface-tension gradient that pulls liquid along the surface (Marangoni flow), generating propulsion for tiny "soap boats" without any moving parts.

FACT 10

FIRE BECOMES A BLUE SPHERE IN MICROGRAVITY

Without buoyant convection, flames aren't teardrops. Combustion becomes diffusion-dominated, producing small, dimmer, often bluish spherical flames that burn differently because oxygen transport is fundamentally altered.

FACT 11

POLARIZED LIGHT REVEALS STRESS INSIDE PLASTICS

Many transparent plastics become birefringent under stress. Viewed between polarizers, they show colored fringe patterns that map stress differences—turning invisible internal forces into a visible "stress fingerprint."

FACT 12

COLD TOUCH DEPENDS ON HEAT EFFUSIVITY

Metal and wood at the same temperature feel different because your skin senses heat flow, not temperature alone. High-effusivity materials draw heat from your skin faster, producing a stronger “cold” sensation.

FACT 13

MICROWAVES HEAT WITHOUT A SPECIAL RESONANCE

Microwave ovens don't rely on a unique "water resonance." Heating comes from dielectric loss: oscillating fields make polar molecules rotate and dissipate energy. Operating near 2.45 GHz also aligns with ISM spectrum allocations.

PART II
THE HUMAN BODY AND BRAIN

FACT 14

SLEEP BOOSTS THE BRAIN'S WASTE CLEARANCE

Experiments show sleep increases fluid exchange between cerebrospinal and interstitial compartments, accelerating removal of metabolites compared with wakefulness. It reframes sleep as an active maintenance mode, not inactivity.

FACT 15

THE BRAIN HAS LYMPHATIC DRAINAGE ROUTES

Functional lymphatic vessels in the meninges drain fluid and immune cells toward deep cervical lymph nodes. This reshaped the long-held “no lymphatics in the brain” assumption in neuroimmunology.

FACT 16

PUPILS WIDEN WITH MENTAL EFFORT

Pupil dilation tracks cognitive control demands: harder tasks reliably increase pupil size, reflecting shifts in attention/arousal systems. That's why "pupillometry" is used as a noninvasive effort proxy in lab studies.

FACT 17

YOU'RE ROUGHLY ONE-TO-ONE WITH MICROBES

The popular "10 bacteria per human cell" claim is outdated. A careful reassessment estimates bacterial and human cell counts are of the same order, closer to roughly 1:1 in a reference adult.

FACT 18

INNATE IMMUNITY CAN FORM TRAINED MEMORY

Innate immune cells can respond more strongly to later challenges after a first exposure, via lasting metabolic and epigenetic changes—an immune “memory” that doesn’t depend on antibodies or T-cell specificity.

FACT 19

BONES ACT LIKE AN ENDOCRINE ORGAN

Bone is not just structural. Bone-derived hormones, notably osteocalcin, participate in systemic regulation including energy metabolism, integrating skeleton, pancreas, and other organs into endocrine feedback loops.

FACT 20

A BONE HORMONE CAN SHAPE MEMORY

Evidence from animal work links osteocalcin to brain development and function: deficits in osteocalcin signaling can impair learning/memory and alter anxiety-like behavior, implying a skeleton-to-brain hormonal channel.

FACT 21

HEARTBEATS CAN AMPLIFY THREAT PERCEPTION

Perception isn't continuous: the timing of a stimulus within a cardiac cycle matters. Fearful faces are detected more easily and judged more intense when presented during systole versus diastole.

FACT 22

YOUR GUT CARRIES TASTE SENSORS TOO

Sweet taste receptors aren't confined to the tongue. Similar receptor machinery in the gut contributes to "nutrient sensing," including incretin hormone secretion and downstream effects on glucose regulation.

FACT 23

PREGNANCY LEAVES BEHIND LONG-LIVED FETAL CELLS

Fetal microchimerism describes fetal-origin cells that can persist in maternal blood and tissues for decades after pregnancy. Their health effects may be beneficial (repair) or harmful (immune conflict).

FACT 24

CIRCADIAN CLOCKS RESHAPE DRUG METABOLISM TIMING

Drug absorption, enzyme activity, and toxicity can vary by time of day because drug-metabolizing pathways are under circadian control. “When” you dose can meaningfully alter pharmacokinetics and outcomes.

FACT 25

THE LIVER CAN REGROW LOST MASS

After surgical resection, the liver undergoes compensatory regeneration. Human data show liver mass restoration can be substantially complete within months, reflecting a rare organ-level "size reset" program.

FACT 26

BRAIN TISSUE ITSELF LACKS PAIN SENSORS

Brain parenchyma has no nociceptors, which is why awake neurosurgery is possible: pain comes from scalp, skull, and meninges. Surgeons can map functions while sparing speech or motor regions.

PART III
EARTH SYSTEMS AND GEOGRAPHY

FACT 27

SEA LEVEL IS GRAVITY'S BUMPY REFERENCE

"Sea level" isn't a perfect shell. Earth's gravity field and ocean circulation create a lumpy geoid and dynamic sea surface, so "level" varies by location even without tides.

FACT 28

MAGNETIC NORTH DRIFTS FASTER THAN YOU THINK

The magnetic pole migrates because Earth's core dynamics change the geomagnetic field. Modern models track substantial pole motion over recent decades, requiring periodic updates to navigation references.

FACT 29

ATMOSPHERIC RIVERS MOVE OCEANS OF MOISTURE

Long, narrow corridors of intense water-vapor transport can deliver extreme precipitation when they make landfall. They function like literal "rivers in the sky," moving moisture thousands of kilometers.

FACT 30

BIG QUAKES NUDGE EARTH'S SPIN RATE

Large earthquakes can redistribute mass and slightly change Earth's rotation and axis, altering day length by microseconds. Tiny changes, but measurable and physically real at planetary scale.

FACT 31

EARTH'S LARGEST DESERT SITS AT POLES

Deserts are defined by low precipitation, not heat. Polar regions can be "deserts" because cold air holds little moisture and snowfall is minimal, despite massive ice on the ground.

FACT 32

CONTINENTS DRIFT ABOUT AS FINGERNAILS GROW

Plate tectonics moves continents centimeters per year—comparable to fingernail growth. Space geodesy (e.g., GPS measurements) makes this drift directly measurable in human lifetimes.

FACT 33

EARTH'S AXIS SLOWLY TRACES A CIRCLE

Earth's rotation axis wobbles (precession) on ~26,000-year timescales, shifting which stars mark "north" and subtly affecting seasonal solar geometry—one component in long-term climate forcing.

FACT 34

OCEANS HIDE MOST OF THE ADDED HEAT

The ocean absorbs the vast majority of excess heat trapped by greenhouse forcing, making ocean heat content a primary indicator of long-term energy imbalance in the climate system.

FACT 35

SATELLITES WEIGH EARTH BY TRACKING GRAVITY

Gravity-measuring satellite pairs detect changes in mass distribution by monitoring tiny variations in their separation. This allows tracking groundwater, ice loss, and ocean mass shifts from space.

FACT 36

SOME DUNES SING DURING SAND AVALANCHES

“Booming dunes” produce sustained tones when sand flows. The sound depends on grain properties and collective motion, turning an avalanche into a natural musical instrument under the right conditions.

FACT 37

VOLCANOES EXHALE MOSTLY WATER, CO_2, SO_2

Volcanic gases are dominated by water vapor, with major contributions from carbon dioxide and sulfur dioxide. These emissions can impact health locally and climate indirectly via aerosols and chemistry.

FACT 38

OCEAN CURRENTS REDISTRIBUTE HEAT AND MOISTURE

The ocean influences climate by storing solar energy and transporting heat and moisture around the planet via currents, shaping regional weather patterns and long-term climate behavior.

FACT 39

AIR PRESSURE IS THE WEIGHT ABOVE

Atmospheric pressure falls with altitude because there is less air mass overhead. It's literally the weight of the atmosphere per unit area, decreasing as the "air column" shortens.

PART IV
BIOLOGY AND NATURAL NETWORKS

FACT 40

SLIME MOLDS DESIGN SURPRISINGLY EFFICIENT NETWORKS

A slime mold can grow transport networks that approximate efficient tradeoffs between resilience and path length, and experiments have shown its networks can mirror human-designed infrastructure layouts.

FACT 41

TREES CAN TRADE CARBON THROUGH FUNGAL LINKS

Experiments demonstrate carbon transfer between plants via mycorrhizal networks. The ecological meaning can be context-dependent and debated, but cross-plant transfer itself has empirical support.

FACT 42

BEES MAP DISTANCE AND DIRECTION BY DANCING

The waggle dance encodes direction relative to the Sun and approximates distance to food sources. It's a symbolic communication system that recruits nestmates using a map-like bodily code.

FACT 43

SOME BACTERIA NAVIGATE USING INTERNAL COMPASS CRYSTALS

Magnetotactic bacteria build magnetosome chains—intracellular magnetic particles—that align cells with Earth's magnetic field, helping them orient in gradients such as oxygen or redox conditions.

FACT 44

STRESSED PLANTS EMIT ULTRASONIC CLICKS INTO AIR

Experiments report that plants under stress (e.g., dehydration or cutting) can emit ultrasonic airborne sounds. These cues are not just measurable; they may carry ecological information in some contexts.

FACT 45

TINY CORAL POLYPS BUILD STONE REEFS

Coral reefs are built by countless small polyps that secrete calcium carbonate skeletons. Over time, living tissue grows atop accumulated mineral structure, creating massive habitats from microscopic labor.

FACT 46

BACTERIA COORDINATE AS SWARMS VIA QUORUM SIGNALS

Many bacteria use quorum sensing: chemical signaling that changes gene expression when population density passes thresholds. Individual cells behave differently once the group is "large enough" to matter.

FACT 47

ANT COLONIES SELF-ORGANIZE WITHOUT ANY LEADER

Ant colonies show distributed control: local interactions and simple rules can yield collective foraging, nest choice, and task allocation without centralized command—an emergent "colony mind."

FACT 48

DEEP-SEA LIFE GLOWS MORE OFTEN THAN NOT

Surveys suggest a large fraction of deep-sea animals produce their own light. Bioluminescence is not an exception in the deep ocean; it is a dominant biological “language” there.

FACT 49

LICHENS ARE TWO ORGANISMS IN ONE

A lichen is not a single organism but a stable symbiosis: a fungus plus a photosynthetic partner (algae and/or cyanobacteria). Together they form a structure neither partner makes alone.

FACT 50

NIGHT-OPENING PLANTS SAVE WATER WITH CAM

CAM photosynthesis shifts gas exchange to nighttime: stomata open when humidity is higher and temperatures lower, reducing water loss while storing CO_2 for daytime sugar production.

FACT 51

SOIL MICROBES MAKE NITROGEN USABLE FOR LIFE

Most organisms can't use atmospheric N_2 directly. Biological nitrogen fixation—performed by microbes—converts inert N_2 into bioavailable forms, underpinning ecosystems and human food systems.

FACT 52

INSECTS EAVESDROP ON PLANT DISTRESS SOUNDS

Recent experiments show some insects can use ultrasonic emissions from stressed plants in decision-making (e.g., where to lay eggs), suggesting an unexpected acoustic channel in plant–insect interactions.

PART V
TECHNOLOGY AND ENGINEERING

FACT 53

QR CODES REBUILD DATA FROM MISSING PIECES

QR codes intentionally add redundancy. With error correction, scanners can recover the message even when portions are obscured or damaged, trading a denser pattern for robustness in the physical world.

FACT 54

REED–SOLOMON CODES MAKE ERRORS MENDABLE

Reed–Solomon coding treats messages as polynomials over finite fields, enabling correction of burst errors. The math is abstract, but its practical impact is everywhere from storage to communication.

FACT 55

COMPACT DISCS SURVIVE SCRATCHES VIA INTERLEAVING

CDs use cross-interleaved error correction so localized scratches become distributed “small” errors after deinterleaving, allowing the decoder to recover audio data that would otherwise be lost.

FACT 56

NOISE SETS A HARD INFORMATION SPEED LIMIT

Shannon showed that for a noisy channel there is a maximum reliable information rate (capacity). Below that rate, error-correcting codes can make communication arbitrarily reliable in principle.

FACT 57

BATTERY LIFE DEPENDS ON A NANO-SCALE FILM

Lithium batteries form a “solid electrolyte interphase” (SEI): electronically insulating yet ion-conducting. It stabilizes the anode but costs lithium/electrolyte and heavily influences lifetime and safety.

FACT 58

FIN-SHAPED TRANSISTORS FIGHT LEAKAGE TODAY

FinFETs wrap the gate around a thin “fin,” improving electrostatic control and reducing short-channel leakage as transistors shrink. This 3D geometry enabled continued scaling when planar control weakened.

FACT 59

HARD DRIVES READ BITS WITH GIANT MAGNETORESISTANCE

Giant magnetoresistance enabled highly sensitive read heads, supporting drastic miniaturization of magnetic storage. It’s a nanoscale quantum effect that made everyday high-capacity drives possible.

FACT 60

BLUE LEDS UNLOCKED EFFICIENT WHITE LIGHTING

Efficient blue LEDs made high-quality white LED lighting feasible (blue + phosphor conversion). The result is a major leap in lighting efficiency and durability compared with incandescent technologies.

FACT 61

FIBER AMPLIFIERS LET LIGHT TRAVEL FARTHER

Erbium-doped fiber amplifiers boost optical signals directly in fiber, enabling long-haul networks with fewer electronic conversions. They are foundational to modern high-capacity optical communication infrastructure.

FACT 62

INTERNET TIME IS A DISCIPLINED GROUP ESTIMATE

Network Time Protocol doesn't trust one clock. It combines multiple time sources and filters out outliers to steer local clocks, creating a robust consensus time across unreliable networks.

FACT 63

MEMORY ALLOYS SNAP BACK WHEN WARMED

Shape-memory alloys can recover a trained shape after deformation when heated through a phase transformation. It looks like "metal remembering," but it's reversible crystal-structure switching.

FACT 64

TIN DROPLETS CREATE LIGHT FOR TINY CHIPS

Extreme ultraviolet lithography generates EUV by striking fast tin droplets with powerful lasers, creating a plasma that emits EUV light—turning molten metal and laser pulses into a chip-printing strobe.

FACT 65

GPS WOULD DRIFT QUICKLY WITHOUT RELATIVITY

Satellite navigation depends on precise timing. Relativistic effects measurably shift satellite clock rates; without continuous correction, location errors would accumulate rapidly and make accurate positioning impossible.

PART VI
TIME, SPACE AND THE UNIVERSE

FACT 66

GPS NEEDS RELATIVITY CORRECTIONS EVERY DAY

Satellites experience weaker gravity and high velocity, shifting clock rates versus Earth clocks. The net effect is tens of microseconds per day—small in time, huge in distance without correction.

FACT 67

GRAVITY SHIFTS YOUR CLOCK OVER 33 CM

Ultra-precise optical clocks measured gravitational time dilation for height changes of mere tens of centimeters. Relativity isn't just for black holes; it's detectable on laboratory scales.

FACT 68

RELATIVITY SHOWS UP AT WALKING SPEEDS

Using optical clocks, researchers observed time dilation from relative speeds under ~10 m/s. It's tiny, but measurable: special relativity is present even in everyday human motion.

FACT 69

THE UNIVERSE GLOWS AT 2.725 KELVIN

The cosmic microwave background is measured with extraordinary precision: a near-perfect relic radiation field that sets a “temperature of space” at about 2.725 K today.

FACT 70

OBSERVABLE UNIVERSE SPANS 93 BILLION LIGHT-YEARS

Because space expands while light travels, the "now-distance" to the most distant observable regions exceeds the universe's age in light-years. The observable diameter is about 93 billion light-years.

FACT 71

MOST OF THE COSMOS IS DARK COMPONENTS

Precision cosmology supports a universe where ordinary matter is a small fraction: most mass-energy is attributed to dark matter and dark energy in the standard ΛCDM framework.

FACT 72

SPACETIME RIPPLES WERE DETECTED IN 2015

A landmark detection captured gravitational waves from a binary black hole merger, confirming a key prediction of general relativity and opening gravitational-wave astronomy as an observational discipline.

FACT 73

ORPHAN PLANETS REVEAL THEMSELVES BY MICROLENSING

Free-floating or wide-orbit planets can be detected when their gravity briefly magnifies background starlight. Microlensing catches planets by their mass, not reflected or emitted light.

FACT 74

STARQUAKES LET US WEIGH AND AGE STARS

Asteroseismology reads oscillation frequencies in starlight. Those "ringing modes" encode density and structure, enabling estimates of fundamental parameters like mass, radius, and age without direct sampling.

FACT 75

INTERSTELLAR SPACE AVERAGES ONE ATOM PER CC

The “vacuum” between stars is not empty, but it is extremely sparse by Earth standards—often around one atom per cubic centimeter on average, with denser clouds still remarkably thin.

FACT 76

Cosmic rays can trigger eye flashes

Astronauts frequently report light flashes in darkness. Reviews link these perceptions to interactions between high-energy cosmic radiation and the visual system, turning particle physics into subjective experience.

FACT 77

SOLAR NEUTRINOS STREAM THROUGH YOU UNNOTICED

Solar fusion produces an enormous neutrino flux; on the order of tens of billions cross each square centimeter of your body per second, yet almost all pass through without interacting.

FACT 78

TWO METHODS DISAGREE ON EXPANSION RATE

Measurements of the Hubble constant from early-universe probes versus local distance ladders can differ beyond expected uncertainties. This "H_0 tension" is a major open problem in modern cosmology.

PART VII
HUMAN BEHAVIOR AND SOCIETY

FACT 79

LOSSES STING MORE THAN EQUAL GAINS

People tend to weigh losses more heavily than mathematically equivalent gains, a core feature of prospect theory. This asymmetry helps explain risk aversion, inertia, and why "breaking even" feels special.

FACT 80

CROWDS CAN REDUCE HELPING, ESPECIALLY UNCERTAINTY

Meta-analysis supports a bystander effect: more bystanders often reduces intervention. The effect weakens in dangerous emergencies, where clarity and urgency can counter diffusion of responsibility.

FACT 81

FAMILIARITY CAN CREATE LIKING WITHOUT REASONS

Mere repeated exposure to a stimulus can increase positive evaluations even without new information. The mind often translates processing fluency into a feeling of preference.

FACT 82

EMOTIONS SPREAD ONLINE WITHOUT DIRECT CONTACT

A large-scale experiment manipulating social-media feed content found measurable “emotional contagion”: exposure to more positive (or negative) posts influenced users’ expressed emotions.

FACT 83

WE OVERESTIMATE HOW LONG FEELINGS LAST

People routinely mispredict their future emotional reactions. A common error is the "impact bias": overestimating the intensity and duration of feelings after future events.

FACT 84

ENDINGS DOMINATE HOW EXPERIENCES ARE REMEMBERED

Memories of experiences overweight the most intense moment and the ending, not the total sum. Adding a "better end" can make a longer unpleasant episode remembered as less bad.

FACT 85

NORM FEEDBACK CAN BACKFIRE WITHOUT APPROVAL

Descriptive norms ("most people do X") can cause a boomerang effect: low consumers increase usage. Adding injunctive cues (approval/disapproval) can fix the backfire and steer behavior.

FACT 86

FRIEND CIRCLES FORM LAYERS, NOT ONE LIMIT

The famous “about 150 relationships” idea is influential but contested. Evidence suggests layered social circles exist, but strict numeric limits vary with context and measurement choices.

FACT 87

PEOPLE PAY TO PUNISH UNFAIR OFFERS

In ultimatum bargaining, people often reject low offers even at personal cost, and cross-cultural data show systematic deviations from pure self-interest across many societies.

FACT 88

SIMILARITY BREEDS CONNECTION IN SOCIAL NETWORKS

Homophily is a near-universal network pattern: ties form disproportionately among similar individuals. It shapes friendship, marriage, work ties, and information flow—often amplifying social clustering.

FACT 89

TEAMS SHOW A MEASURABLE COLLECTIVE INTELLIGENCE

Groups exhibit a “c factor” predicting performance across tasks. It correlates weakly with max or average IQ, but more with social sensitivity and balanced conversational turn-taking.

FACT 90

WORDS HIJACK COLOR NAMING IN STROOP

Reading is so practiced it intrudes automatically. Naming ink color becomes slower and error-prone when the printed word conflicts, exposing limits of "willpower" over habitual processing.

FACT 91

EQUAL TURN-TAKING PREDICTS SMARTER GROUP DECISIONS

Groups perform better when participation is distributed rather than dominated by a few voices. Conversational equality is repeatedly linked to group-level performance across varied collaborative problem types.

PART VIII
HIDDEN PATTERNS AND SYSTEMS

FACT 92

MOST NUMBERS START WITH 1, NOT 9

Many real-world datasets have leading digits that follow a logarithmic pattern rather than uniform chance. This "first-digit law" was noted in the 19th century and quantified with large empirical tables later.

FACT 93

BENFORD PERSISTS WHEN QUANTITIES CHANGE UNITS

A striking property of first-digit distributions is scale invariance: converting meters to feet often doesn't change the leading-digit pattern for many broad, multi-scale datasets.

FACT 94

A FEW SHORTCUTS MAKE NETWORKS SMALL-WORLD

Networks can have high clustering yet very short average path lengths if only a small fraction of long-range links exist. Tiny rewiring can collapse "degrees of separation."

FACT 95

HUBS FORM WHEN LINKS PREFER THE POPULAR

Preferential attachment—new nodes linking more often to already well-connected nodes—naturally generates hubs and power-law-like degree distributions. “The rich get richer” becomes network geometry.

FACT 96

LOG–LOG PLOTS CAN FOOL YOUR EYES

Many datasets look straight on log–log axes, but that's not enough to prove a power law. Rigorous fitting and model comparison are needed; naive regression often produces false certainty.

FACT 97

AVERAGES CAN REVERSE THE APPARENT RELATIONSHIP

Simpson’s paradox occurs when a trend in multiple groups reverses after aggregation. The “paradox” is a warning: association without causal structure can be profoundly misleading.

FACT 98

CONNECTIVITY APPEARS SUDDENLY AT A THRESHOLD

In percolation theory, as link probability increases, disconnected clusters abruptly form a spanning "giant component" at a critical threshold—an archetypal phase transition in networks and materials.

FACT 99

COASTLINES GROW LONGER AS RULERS SHRINK

Measured coastline length depends on the measurement scale: smaller rulers capture more detail, increasing total length. Fractal geometry formalizes this with non-integer "dimension."

FACT 100

DELAYS CAN TURN FEEDBACK INTO OSCILLATIONS

In feedback systems, delays can destabilize control: corrective actions arrive late, overshoot, and create oscillations. The same logic appears from engineering control loops to biological regulation.

FACT 101

METRICS BREAK WHEN THEY BECOME TARGETS

When a quantitative measure becomes a performance target, behavior adapts to the metric, corrupting its informational value. It's a systems warning: optimization can destroy signal.

FACT 102

RANKED THINGS OFTEN FOLLOW ZIPF-LIKE SLOPES

In many collections, the frequency of the r-th ranked item scales roughly like 1/r. This rank-frequency regularity appears across language, cities, and other complex systems.

FACT 103

Rare events can dominate averages

In heavy-tailed (fat-tailed) distributions, “typical” averages can be misleading because extreme events contribute a large share of totals. Stability requires careful thinking about tails and sampling.

FACT 104

A FEW CREATORS PRODUCE MOST OUTPUT

Scientific productivity is strongly skewed: a small fraction of authors produce a large fraction of papers. This “Lotka-like” concentration pattern recurs across creative and economic outputs.

CONCLUSIONS

We've traveled together through a curious landscape where science meets everyday life, where technology touches psychology and nature interacts with culture. The facts you've read are not just isolated pieces of information, but small lenses through which to observe the world. Some talk about invisible particles, others about brains, maps or music. Yet all of them tell the same story: reality is stranger, more interconnected and more surprising than it appears. We live immersed in systems that react, adapt and synchronize, often without us even noticing. This book was not designed to give you final answers, but to train your perception. To help you notice what usually goes unseen. Because understanding even a little more about how things work means living in a more conscious, creative and free way.

Thinking in Connections

One of the most important messages that emerges from these hundred-plus facts is that almost nothing exists in isolation. The brain talks to the gut. Music can change your heart rate. Light can influence crime. Continents change shape on maps depending on how we choose to represent them. Even two particles separated by enormous distances can remain linked in ways that challenge intuition. This means that the world is not a sum of independent objects, but a network of dynamic relationships.

Thinking in connections means stopping the search for single causes and starting to observe systems. It means understanding that a food choice can influence mood, that the design of a city can change social behavior, that an algorithm can reshape global economics. Every phenomenon is part of a chain of effects, often invisible, that propagate through time and space.

Today, genius is not just about knowing many things, but about connecting them. Seeing analogies across different fields. Recognizing patterns where others see only randomness. This is the kind of intelligence required in a complex world: the ability to move between different levels of reality, from the microscopic to the social, from the biological to the digital.

When you begin to think in terms of connections, the way you make decisions also changes. You become more aware of indirect consequences, more sensitive to feedback loops, more open to the possibility that small variations can produce large effects. In other words, you develop a form of systemic thinking that allows you to navigate uncertainty more effectively.

Many of the facts you've read may seem like curiosities. But they are actually examples of the same underlying logic: interdependence. And learning to recognize it is one of the first steps toward understanding how the modern world truly works.

Training the Curious Eye

Knowledge does not arise only from study, but from attention. From noticing that something doesn't quite add up. From asking why apples float, why skyscrapers sway, why perfume changes over time or why a coincidence feels meaningful. Curiosity is a form of operational intelligence. It's what transforms experience into learning.

Training the curious eye means developing the habit of asking questions. Not only big philosophical ones, but also small practical ones. How does this work? Why does this happen? What would change if...? This continuous questioning is what allows unexpected connections to emerge and new ideas to take shape.

In an age where information is everywhere, the real resource is not access to data, but the ability to interpret it. To select it. To place it within a meaningful framework. Facts alone are not enough. They must be observed, compared and connected. Only then do they become tools for understanding.

Curiosity also has an emotional dimension. It keeps wonder alive, reduces mental rigidity and encourages adaptation. Those who remain curious tend to explore more, experiment more and shift perspectives more easily. And this is essential in a world that evolves rapidly.

Perhaps the most important outcome of this journey is not what you have learned, but how you have learned to look. If you now find yourself pausing in front of an everyday phenomenon and wondering what makes it possible, then this book has achieved its purpose. Because every new question is the beginning of a new discovery.

ABOUT THE AUTHOR

The Top Tier Education Program is a publishing initiative dedicated to designing and producing advanced strategic books for entrepreneurs, managers, professionals, and individuals seeking tools to navigate contemporary transformation. The Program focuses on developing content that combines theoretical depth and practical applicability, leveraging a transdisciplinary and systemic approach.

Each volume is designed as a structured learning platform, offering conceptual maps and toolkits to orient readers in challenges such as leadership, strategy, innovation, sustainability, and organizational culture. The mission is to deliver high-quality, accessible resources that empower readers to understand the hidden structures shaping the emerging world order, without making claims regarding the individual authorship of the content.

www.ingramcontent.com/pod-product-compliance
Lightning Source LLC
LaVergne TN
LVHW022227200926
848481LV00006B/2152

* 9 7 9 1 2 2 4 3 1 4 3 2 5 *